WEIRD ANIMAL SCIENCE

ANIMAL MYSTERIES

THE BREAKDOWN

10877506

VIRGINIA LOH-HAGAN

45TH PARALLEL PRESS

Published in the United States of America by Cherry Lake Publishing Group
Ann Arbor, Michigan
www.cherrylakepublishing.com

Photo Credits: © Cascadia Research Collective; NOAA News, cover, 1; © Svetsol/Shutterstock, cover, 1; © bsd studio/Shutterstock, cover, 1, 20; © Malenkka/Shutterstock, cover, 1; © m-kojot/iStock, 7; © f8grapher/Shutterstock, 8; © ratna sriyuli/Shutterstock, 10; © Friendly Studios/Shutterstock, 11; © Andy333/Shutterstock, 12; © nikiteev_konstantin/Shutterstock, 13; © Ajit S N/Shutterstock, 14; © Wirestock/iStock, 17; © Gerald Corsi /iStock, 18; © National Marine Sanctuaries, Public domain, via Wikimedia Commons, 21

Graphic Element Credits: Graphic Element Credits: Cover, multiple interior pages: © WindAwake/Shutterstock; © Theo_hrm/Shutterstock; © Rodin Anton/Shutterstock; © Petr Vaclavek/Shutterstock; © Nikolaeva/Shutterstock

Copyright © 2025 by Cherry Lake Publishing Group

All rights reserved. No part of this book may be reproduced or utilized in any form or by any means without written permission from the publisher.

45th Parallel Press is an imprint of Cherry Lake Publishing Group.

Library of Congress Cataloging-in-Publication Data has been filed and is available at catalog.loc.gov

Cherry Lake Publishing Group would like to acknowledge the work of the Partnership for 21st Century Learning, a Network of Battelle for Kids. Please visite Battelle for Kids online for more information.

Printed in the United States of America

Note from publisher: Websites change regularly, and their future contents are outside of our control. Supervise children when conducting any recommended online searches for extended learning opportunities.

Dr. Virginia Loh-Hagan is an author, university professor, and former classroom teacher. She's currently the Director of the Asian Pacific Islander Desi American Resource Center at San Diego State University. She loves science-fiction stories. She lives in San Diego with her very tall husband and very naughty dogs.

TABLE OF CONTENTS

Break It Down 4

Test It Out 5

Animal Mysteries 6

Put It Together 22

Learn More 23

Glossary 24

Index 24

BREAK IT DOWN

1. DECODE IT

- Look at each letter.
- Look at letter patterns.
- Blend sounds.
- Try other possible sounds.

2. BUILD VOCABULARY

- Listen to the word.
- Match the sounds to a word you know.
- Look for context clues for words you don't know.
- Look up the meaning in a glossary or dictionary.

3. MAKE CONNECTIONS

- Name other examples.
- Draw the idea.
- Compare ideas.
- Ask questions.

TEST IT OUT

WORDS YOU'LL LEARN

Look over the list of vocabulary words. Are there any you already know?

fingerprints
krill
reproduction
vertically
whorlz

GET STARTED

Familiar word parts are usually pronounced the same everywhere. Examples include smaller words in compound words and suffixes like *-ly* and *-tion*.

- Look at the letters.
- Chunk long words to read them.
- Look for familiar word parts.

The word *fingerprints* is made of 2 smaller words: *finger* and *prints*. It is pronounced: FING-er-prints.

- The 1st vowel is a short *i*: *fing-*.
- The 2nd vowel has an *r* after it. It says "er": *-er*.
- The 3rd vowel is a short *i*: *-prints*.

Scientists study animal mysteries. One mystery is in Australia. Koalas live there.

BUILD VOCABULARY

Whorls (WUHRLZ) are lines that move out in a circle from one point.

- Whorls are like swirls.
- They are in fingerprints. There can be whorls in water.
- Other forms: whorled, whorlflower, whorlywort

They have **fingerprints** like humans. Fingerprints are patterns of ridges on fingertips.

The patterns have loops, **whorls**, and arches.

Some scientists think fingerprints are connected to grasping.

Koalas climb **vertically**.

That means they move up and down.

DECODE IT

The word *vertically* has 4 vowel sounds. It has a familiar word part: *-ly*. This suffix is pronounced "lee."

- The 1st vowel is an *r*-controlled *e* and says "*er*."
- The 2nd vowel is a short *i*.
- The 3rd vowel says "uh."
- The last vowel is a long *e*.
- *Vertically* is pronounced VER-tih-kuh-lee.

They reach out. They grab leaves.

They bring them to their mouths.

The fingerprints may be needed for fine motor control.

Chimps and gorillas also have fingerprints.

They're related to humans.

But koalas are not related to humans!

That's why it's a mystery!

Another mystery is deep in the ocean. Blue whales are the world's largest animals.

They're longer than 2 school buses.

They weigh as much as 33 elephants.

Their hearts are the size of cars.

They eat about 4 tons of **krill** a day.

Krill are like shrimp.

MAKE CONNECTIONS

A human eats about 3 to 5 pounds (1.4 to 2.3 kilograms) of food per day. A whale eats 4 tons. That's the same as 8,000 lb. (3,628.7 kg)!

- What would happen if there weren't enough krill for these whales to eat?

Blue whales are also the world's loudest animals.

They're louder than a jet engine!

Yet, blue whales are a mystery. Not much is known about them.

Scientists don't know how long they live.

They don't know their **reproduction** behaviors.

Reproduction means making babies.

DECODE IT

The word *reproduction* has 4 vowel sounds. It has a familiar word part: *-tion*. This suffix is pronounced "shuhn."

- Break it into chunks: re-pro-duc-tion.
- The 1st vowel sound is a long *e*.
- The next 3 vowel sounds say "uh."
- *Reproduction* is pronounced ree-pruh-DUK-shuhn.

It’s hard to study them. First, there aren’t many of them to study.

Second, they’re shy animals. They don’t come to the surface often.

Lastly, they travel long distances all over the ocean. This makes them hard to track.

There's so much left to learn about animals! Maybe you will solve these animal mysteries!

PUT IT TOGETHER

Main Idea:
There is still a lot that scientists don't know about animals.

Detail:
Koalas may use their fingerprints to grip and climb.

Detail:
Koalas are not related to humans, but still have fingerprints.

Detail:
Blue whales are the world's largest animals. Scientists don't know how long they live.

Detail:
Blue whales are hard to study.

Detail:
Scientists don't know about blue whale reproduction.

LEARN MORE

BOOKS

Jenkins, Steve. *The Animals Book*. Boston, MA: HMH Books for Young Readers, 2013.

Lerwill, Ben, and Sarah Walsh (illus.). *WildLives: 50 Extraordinary Animals That Made History*. New York, NY: Atheneum Books for Young Readers, 2020.

Watson, Galadriel, and Samantha Dixon (illus.) *Running Wild: Awesome Animals in Motion*. Toronto, ON: Annick Press, 2020.

ONLINE

Search these online sources with an adult:

- National Geographic Kids | Animals
- Britannica Kids | Animal Kingdom

GLOSSARY

krill (KRIL) shrimp-like crustaceans; a blue whale's main food source

mating (MAYT-ing) reproducing and making babies

reproduction (ree-pruh-DUK-shuhn) the act or process of producing young

vertically (VUHR-tih-kuhl-ee) up and down

whorls (WUHRLZ) lines that move out in a circle from one point

INDEX

blue whales, 14–22

chimps, 12

elephants, 15

fingerprints, 7–8, 10, 12, 22

gorillas, 12

koalas, 6–11, 13, 22

krill, 16

mysteries, 6, 14, 18

reproduction, 19

shrimp, 16

vocabulary, 4–6